CPSIA information can be obtained
at www.ICGtesting.com
Printed in the USA
LVHW020301160323
741737LV00013B/874

9 783749 707799